Quantitative Methods

With Python

Dr. Abhinandan H. Patil

To my Family members, well wishers and Teachers in that order.

There is no alternative to hard-work and perseverance

Dr. Abhinandan H. Patil

TABLE OF CONTENT

Chapter 1 Crash Course in Relevant Mathematics 8
Chapter 2 Crash Course in Relevant Python 9
Chapter 3 Introduction 10
Chapter 4 Probability and Related Problems 25
Chapter 5 Statistics 42
Chapter 6 Scatter and Regression Models 43
Chapter 7 Forecasting 51
Chapter 8 Linear Programming and Applications 52
Chapter 9 Introduction to Basic Sympy Functionalities 55
Chapter 10 Plotting with Sympy 73
Chapter 11 Summary 75

Acknowledgement

Author Acknowledges Kubuntu operating system for providing host based system. HP for great hardware. MathCha for great software for creating this Book. Libredraw for creating Diagrams. Python community at large is acknowledged. Particularly Sympy, Scipy and MatPlotLib. Author thanks Microsoft for VSCode.

Author acknowledges GitHub for providing the great framework for maintaining the code.

Preface

We touch few capabilities of Python Quantitaive libraries/packages in this Book. There is a high probability that you will find some one has already documented their own experiences with python in the form of Books, Blogs,YouTube etc. Today the Scipy and Sympy user guides combined is in excess of 5000 pages. That is enormous capability available to everyone to carry out their activity in Quantitative Field with Python. And it is growing by the day. It is these libraries that make Python so Rich!!

About the Author

Dr. Abhinandan H. Patil currently resides in India, Karnataka. Before this, he has worked in Wireless Network Software Organization as Lead Software Engineer for close to a decade. His Research out put is available as Books and Thesis in IJSER, USA. He is Active Researcher in the field of Machine Learning, Deep Learning, Data Science, Artificial Intelligence, Regression Testing applied to Networks, Communication and Internet of Things. He is active contributor of Science, Technology, Engineering and Mathematics. He is currently working on few Undisclosed Books. In the capacity as CTO of organization He carries out Research activity. He has started Blogging recently on Technology and Allied Areas. He is nominated for RULA Research Award in year 2019. He is Adarsh Vidya Saraswati Rashtriya Puraskar Awardee in year 2020. Dr. Abhinandan H. Patil is senior IEEE member since 2013 and is member of Smart Tribe and Cheeky Scientists Association. Dr. Patil is awarded for innovation methodolgy in teaching.

Dr. Abhinandan H. Patil can be visited at https://abhinandanhpatil.info and his personal email ID is Abhinandan_patil_1414@yahoo.com

Chapter 1 Crash Course in Relevant Mathematics

This is a crash course relevant to Mathematics

Essential Companions for Your Learning
1. NCERT text Books for 11th Standard and 12th Standard, published in India 2. Comprehensive Engineering Mathematics by John Bird 3. Engineering Mathematics by Croft, Davison, Hergreaves and Flint 4. Mathematics for Engineers and Scientists Andrei and Alexander 5. Advanced Engineering Mathematics by Erwin Kreyszig 6. Sage for Undergraduates by Gregory V. Bard //This Books is for Understanding what softwares are capable today when it comes to Mathematics.

Chapter 2 Crash Course in Relevant Python

This is a crash course relevant to Python

> **Essential Companions for Your Learning**
>
> 1. Scipy Lecture Notes by Scipy
> 2. SciPy Reference Guide, valuable documentation
> 3. Sympy Docs by Sympy Team around 2500 pages of valuable documentation which actually shows what Sympy is capable of
> 4. Python Documentation by Team Python
> 5. Related Python Books by many more Authors such as Cyrille Rossant, David M. Beazley, Allen B. Downey, Peter Bruce & Andrew Bruce, Joel Grus, Aurelian Geron, Amit Saha, Summerfield, Brown. Everybody offers you something unique to learn.
> 6. Scikit Learn documentation for the Machine Learning enthusiasts.

Chapter 3 Introduction

This is the intoduction chapter. Typically the Quantitative approach involves following steps

1. Understanding the problem
2. Acquiring Data
3. Coming up with model
4. Developing solution with software
5. Testing
6. Analysing the results
7. Implementing the results for problem at hand

Let us take standard example of calculating the break even point of buisness. We know:

Profit = Revenue - Total costs
Profit = Revenue - (Fixed cost + Variable cost)

Buisness breaks even when:
0 = Revenue - (Fixed cost + Variable cost)
Revenue = Fixed cost + Variable cost
Selling Price*unit sold = Fixed cost + variable cost per unit * units sold
(Selling Price - variable cost per unit) * unit sold = Fixed cost

$$\text{unit sold} = \frac{FixedCost}{(Selling\ Price - Variable\ cost\ per\ Unit)}$$

To break even this is the conditon. The equation is **Mathematical model.**

Therefore to break even we need to sell $\frac{FixedCost}{(Selling\ Price - Variable\ cost\ per\ Unit)}$ units

Let us write it in Python.

Program 1 Break Even Point Calculator

```
'''
Break even point calculator for given fixed cost and variable
cost
Enter the required values from the prompt
Ex: Maximum units sold -> 1000
Fixed cost -> 1000
```

```
Variable cost per unit -> 5
'''

import matplotlib.pyplot as plt


if __name__ == '__main__':
    # assume values of x
    print(__doc__)
    print('Enter the maximum units sold')
    maximumUnitsSold = int(input())
    print('Enter fixed cost')
    fixedCost = int(input())
    print('Enter the variable cost per unit')
    variableCostPerUnit = int(input())
    print('Enter the selling price per unit')
    sellingPrice = int(input())
    x_values = range(0, maximumUnitsSold, 1)
    y_values = []
    for x in x_values:
        # calculate the values of the y as list
        y_values.append(variableCostPerUnit*x - fixedCost)
    breakEvenPoint = int(fixedCost/(sellingPrice-variableCostPerUnit))
    stringBreakEvenPoint = str(breakEvenPoint)
    print('The business breaks even at',breakEvenPoint,'units')
    plt.plot(x_values,y_values)
    plotTitle = "The buisness breaks even at "+stringBreakEvenPoint+" units"
    plt.title(plotTitle)
    plt.ylabel('Profit')
    plt.show()
```

```
Break even point calculator for given fixed cost and variable cost
```

```
Enter the required values from the prompt
Ex: Maximum units sold -> 1000
Fixed cost -> 1000
Variable cost per unit -> 5

Enter the maximum units sold
1000
Enter fixed cost
1000
Enter the variable cost per unit
5
Enter the selling price per unit
10
The business breaks even at 200 units
```

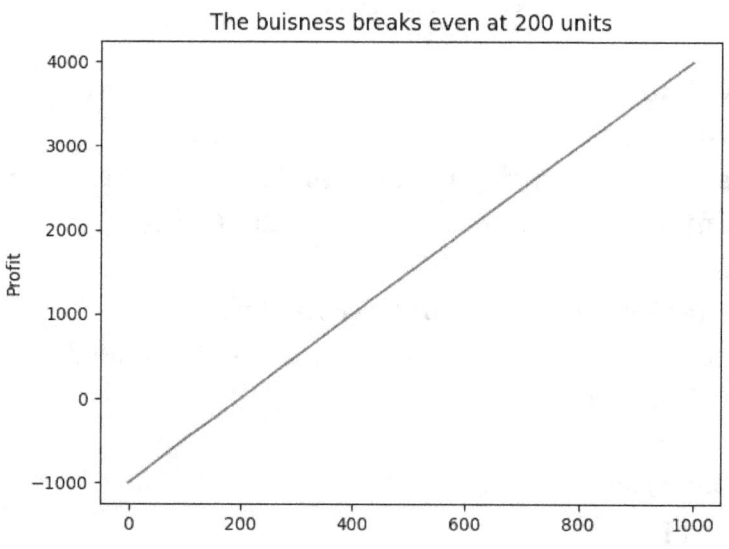

Figure 1: Break Even Point Calculator

When we fail to sell anything we incur loss equal to fixed cost in this case $1000. When we sell 1000 units we make Profit of $4000 and we break even at 200 units.

Program 2 Compounding Factor Calculated Yearly

Let us take an example of computing the compounding factor with compounding **annually**. This is for simplification. Most Banks in several nation compound monthly.

$$\text{Compounding Factor} = \left(1 + \frac{x}{100}\right)^n$$

```
'''
Compounding factor calculator
Calculates the compounding factor for various combination of
interest rate and year based on yearly
compoundng
Inputs none
Input data can be edited in the program
'''

import numpy as np
import matplotlib.pyplot as plt
from mpl_toolkits.mplot3d import Axes3D


fig = plt.figure()
ax = Axes3D(fig)
X = np.arange(4, 10, 1)
print(X)
Y = np.arange(4, 10, 1)
Z = np.zeros((6,6))

print(Y)
for x in X:
    for y in Y:
        print(x,y)
        print(x ,'percent ',y,'years ',(1+x/100)**y)
        Z[x-4][y-4]= (1+x/100)**y
print(Z)
#plt.plot(X,Y,Z)
ax.plot_surface(X,Y,Z)

plt.show()
```

```
[4 5 6 7 8 9]
[4 5 6 7 8 9]
4 percent  4 years  1.1698585600000002
4 percent  5 years  1.2166529024000001
4 percent  6 years  1.2653190184960004
4 percent  7 years  1.3159317792358403
4 percent  8 years  1.3685690504052739
4 percent  9 years  1.423311812421485
5 percent  4 years  1.2155062500000002
5 percent  5 years  1.2762815625000004
5 percent  6 years  1.3400956406250004
5 percent  7 years  1.4071004226562505
5 percent  8 years  1.477455443789063
5 percent  9 years  1.5513282159785162
6 percent  4 years  1.2624769600000003
6 percent  5 years  1.3382255776000003
6 percent  6 years  1.4185191122560004
6 percent  7 years  1.5036302589913606
6 percent  8 years  1.5938480745308423
6 percent  9 years  1.6894789590026928
7 percent  4 years  1.3107960100000002
7 percent  5 years  1.4025517307000004
7 percent  6 years  1.5007303518490005
7 percent  7 years  1.6057814764784306
7 percent  8 years  1.718186179831921
7 percent  9 years  1.8384592124201555
8 percent  4 years  1.3604889600000003
8 percent  5 years  1.4693280768000005
8 percent  6 years  1.5868743229440005
8 percent  7 years  1.7138242687795209
8 percent  8 years  1.8509302102818825
8 percent  9 years  1.9990046271044333
9 percent  4 years  1.4115816100000005
9 percent  5 years  1.5386239549000005
9 percent  6 years  1.6771001108410006
9 percent  7 years  1.8280391208166908
```

```
9 percent   8 years   1.9925626416901934
9 percent   9 years   2.171893279442311
[[1.16985856 1.2166529  1.26531902 1.31593178 1.36856905
  1.42331181]
 [1.21550625 1.27628156 1.34009564 1.40710042 1.47745544
  1.55132822]
 [1.26247696 1.33822558 1.41851911 1.50363026 1.59384807
  1.68947896]
 [1.31079601 1.40255173 1.50073035 1.60578148 1.71818618
  1.83845921]
 [1.36048896 1.46932808 1.58687432 1.71382427 1.85093021
  1.99900463]
 [1.41158161 1.53862395 1.67710011 1.82803912 1.99256264
  2.17189328]]
```

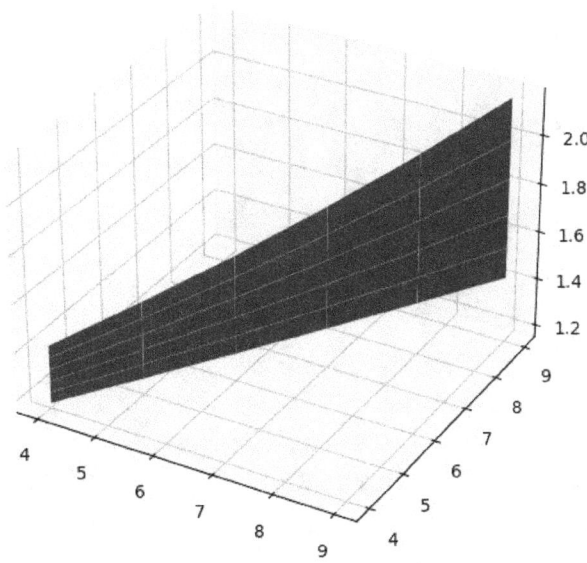

Figure 2: Compounding Factor Calculated Yearly

We get maximum compounding when interest rate is 9% and tenure is 9 years. And lowest with 4% for 4 years. **Obvious**. But if we want exam numbers or quantitaive factor, we need to calculate!

Program 3 Compounding Calculator, Monthly Calculations

Let us refine the above program for monthly compounding.

$$\text{Compounding Factor} = \left(1 + \frac{x}{12*100}\right)^{n*12}$$

```
'''
Compounding factor calculator
Calculates the compounding factor for various combination of
interest rate and year based on monthly
compoundng
Inputs none
Input data can be edited in the program
'''

import numpy as np
import matplotlib.pyplot as plt
from mpl_toolkits.mplot3d import Axes3D


fig = plt.figure()
ax = Axes3D(fig)
X = np.arange(4, 10, 1)
print(X)
Y = np.arange(4, 10, 1)
Z = np.zeros((6,6))

print(Y)
for x in X:
    for y in Y:
        print(x ,'percent ',y,'years ', 'compounding factor: ', (1+x/1200)**(y*12))
        Z[x-4][y-4]= (1+x/1200)**(y*12)

print("Compounding factors in table format")
print(Z)
ax.plot_surface(X,Y,Z)

plt.show()
```

```
[4 5 6 7 8 9]
[4 5 6 7 8 9]
4 percent  4 years  compounding factor:  1.1731986699758588
4 percent  5 years  compounding factor:  1.2209965939421215
4 percent  6 years  compounding factor:  1.2707418790791325
4 percent  7 years  compounding factor:  1.3225138638856104
4 percent  8 years  compounding factor:  1.376395119233124
4 percent  9 years  compounding factor:  1.4324715800579508
5 percent  4 years  compounding factor:  1.2208953550254191
5 percent  5 years  compounding factor:  1.2833586785035118
5 percent  6 years  compounding factor:  1.3490177441587443
5 percent  7 years  compounding factor:  1.4180360522260398
5 percent  8 years  compounding factor:  1.4905854679226442
5 percent  9 years  compounding factor:  1.5668466494164976
6 percent  4 years  compounding factor:  1.2704891610953823
6 percent  5 years  compounding factor:  1.3488501525493075
6 percent  6 years  compounding factor:  1.4320442784916434
6 percent  7 years  compounding factor:  1.520369636082082
6 percent  8 years  compounding factor:  1.6141427084608484
6 percent  9 years  compounding factor:  1.7136994987557481
7 percent  4 years  compounding factor:  1.3220538778853574
7 percent  5 years  compounding factor:  1.4176252596139902
7 percent  6 years  compounding factor:  1.520105504255328
7 percent  7 years  compounding factor:  1.629994054067955
7 percent  8 years  compounding factor:  1.7478264560317114
7 percent  9 years  compounding factor:  1.874176971860912
8 percent  4 years  compounding factor:  1.37566610043379
8 percent  5 years  compounding factor:  1.489845708301605
8 percent  6 years  compounding factor:  1.61350216730999235
8 percent  7 years  compounding factor:  1.7474220514294954
8 percent  8 years  compounding factor:  1.8924572198827105
8 percent  9 years  compounding factor:  2.049530235787287
9 percent  4 years  compounding factor:  1.431405333313711
9 percent  5 years  compounding factor:  1.5656810269415706
9 percent  6 years  compounding factor:  1.7125527068212796
9 percent  7 years  compounding factor:  1.8732019633462298
9 percent  8 years  compounding factor:  2.0489212282389357
9 percent  9 years  compounding factor:  2.241124172232252
Compounding factors in table format
```

```
[[1.17319867 1.22099659 1.27074188 1.32251386 1.37639512
  1.43247158]
 [1.22089536 1.28335868 1.34901774 1.41803605 1.49058547
  1.56684665]
 [1.27048916 1.34885015 1.43204428 1.52036964 1.61414271 1.7136995
  ]
 [1.32205388 1.41762526 1.5201055  1.62999405 1.74782646
  1.87417697]
 [1.3756661  1.48984571 1.61350217 1.74742205 1.89245722
  2.04953024]
 [1.43140533 1.56568103 1.71255271 1.87320196 2.04892123
  2.24112417]]
```

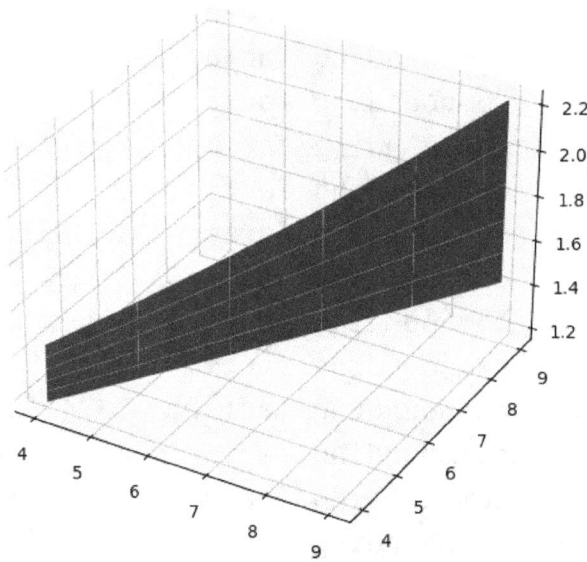

Figure 3: Compounding Factor Calculated Monthly

> **Almost same result. 9 percent 9 years compounding factor: 2.241124172232252 when compounded monthly and 9 percent 9 years 2.171893279442311 when compounded annually.**

Program 4 Example to Show howto Plot Given Revenue, Cost and therefore Profit

Let us take an example where:

$$Revenue(n) = 4.50n - \frac{n^2}{80}$$

$$Cost(n) = 0.50n + 100$$

Therefore

$$Profit = Revenue(n) - Cost(n) = 4n - \frac{n^2}{80} - 100$$

```python
'''
Example to show how to plot given:
Revenue
Cost
and therefore
profit functions graphically for further inference
'''

import matplotlib.pyplot as plt
import math

def revenue(unitsSold):
    return 4.5*unitsSold - ((unitsSold ** 2)/80)

def cost(unitsSold):
    return 0.5*unitsSold + 100

def profit(unitsSold):
    return revenue(unitsSold)-cost(unitsSold)


if __name__ == '__main__':
    # assume values of x
    print(__doc__)
    unitValues = range(0, 350, 1)
    revenueValues = []
    costValues = []
    profitValues = []
```

```
for x in unitValues:
    # calculate the values of the y as list
    revenueValues.append(revenue(x))
    costValues.append(cost(x))
    profitValues.append(profit(x))
plt.plot(unitValues,revenueValues,'-',label='revenue')
plt.plot(unitValues,costValues,'.', label = 'cost')
plt.plot(unitValues,profitValues,':',label = 'profit')
plt.legend()
plt.title("revenue cost profit")
plt.show()
```

Example to show how to plot given:
Revenue
Cost
and therefore
profit functions graphically for further inference

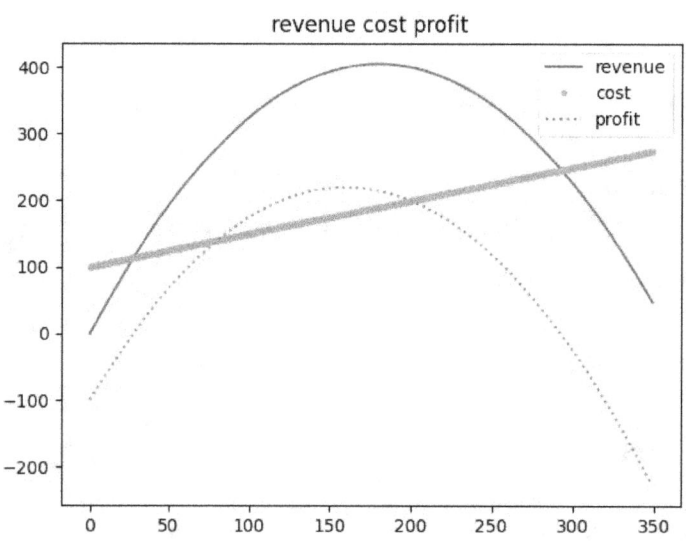

Figure 4: Revenue Cost and Profit for Given Model

> This is how the plot looks like. This can be used for further reference. This problem **definition** is from Sage for undergraduate Book. We have implemented using **Python**

Program 5 EMI Factor Calculator

Now let us take the example of Mathematical model

$$EMI = \left(P * R * \frac{(1+R)^N}{(1+R)^N - 1} \right)$$

```python
'''
EmiFactor calculator
Accepts user input parameters none
Emits tabular EmiFactor for various combination of interest rate
and period
Calculations are monthly
'''

import numpy as np
import matplotlib.pyplot as plt
from mpl_toolkits.mplot3d import Axes3D

def emiFactor(interest, period):
    return ((interest/1200)*(1+interest/1200) ** (period*12))/((1+interest/1200) ** (period*12) -1)
print(__doc__)
fig = plt.figure()
ax = Axes3D(fig)
X = np.arange(4, 10, 1) # Interest rate in percentage
print(X)
Y = np.arange(4, 10, 1) # Number of years
Z = np.zeros((6,6))

print(Y)
for x in X:
    for y in Y:
```

```
        print(x,y)
        print(x ,'percent ',y,'years ', emiFactor(x,y))
        Z[x-4][y-4]= emiFactor(x,y)
print(Z)
#plt.plot(X,Y,Z)
ax.plot_surface(X,Y,Z)

plt.show()
```

EmiFactor calculator
Accepts user input parameters none
Emits tabular EmiFactor for various combination of interest rate and period
Calculations are monthly

[4 5 6 7 8 9]
[4 5 6 7 8 9]
4 percent 4 years 0.022579054641689508
4 percent 5 years 0.018416522055265973
4 percent 6 years 0.01564518307254219
4 percent 7 years 0.013668806336489574
4 percent 8 years 0.012189275302306636
4 percent 9 years 0.011040968903020178
5 percent 4 years 0.023029293570644659
5 percent 5 years 0.018871233644010988
5 percent 6 years 0.016104932661450973
5 percent 7 years 0.014133909071907048
5 percent 8 years 0.012659920012123831
5 percent 9 years 0.011517273168364723
6 percent 4 years 0.023485029047936062
6 percent 5 years 0.019332801529428272
6 percent 6 years 0.0165728878934725
6 percent 7 years 0.014608554483781046
6 percent 8 years 0.013141430210139424
6 percent 9 years 0.012005749630925785
7 percent 4 years 0.023946244662442282
7 percent 5 years 0.019801198540349466

```
7 percent  6 years   0.017049006471969563
7 percent  7 years   0.015092679982189935
7 percent  8 years   0.013633717080503108
7 percent  9 years   0.012506276594370709
8 percent  4 years   0.024412922341502694
8 percent  5 years   0.020276394288841385
8 percent  6 years   0.017533240611952928
8 percent  7 years   0.015586214402669966
8 percent  8 years   0.014136679254544909
8 percent  9 years   0.013018714887236304
9 percent  4 years   0.024885042373934033
9 percent  5 years   0.020758355226353872
9 percent  6 years   0.018025537168271727
9 percent  7 years   0.016089078260038456
9 percent  8 years   0.014650203273692885
9 percent  9 years   0.013542908653137183
[[0.02257905 0.01841652 0.01564518 0.01366881 0.01218928
  0.01104097]
 [0.02302929 0.01887123 0.01610493 0.01413391 0.01265992
  0.01151727]
 [0.02348503 0.0193328  0.01657289 0.01460855 0.01314143
  0.01200575]
 [0.02394624 0.0198012  0.01704901 0.01509268 0.01363372
  0.01250628]
 [0.02441292 0.02027639 0.01753324 0.01558621 0.01413668
  0.01301871]
 [0.02488504 0.02075836 0.01802554 0.01608908 0.0146502
  0.01354291]]
```

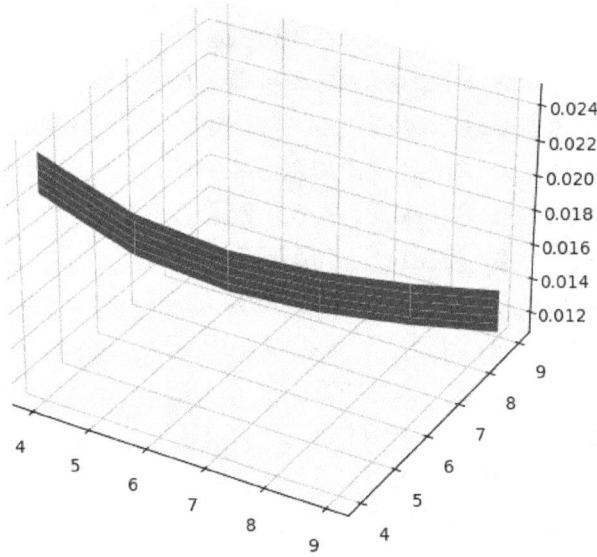

Figure 5: EmiFactorCalculator

> When Banks increase the floating rate interest, try increasing the tenure to keep the EMI in control. On the flip side, long commitment.

So as we mentioned, in all these examples, following steps are involved.

1. Understanding the problem
2. Acquiring Data
3. Coming up with model
4. Developing solution with software
5. Testing
6. Analysing the results
7. Implementing the results for problem at hand

Chapter 4 Probability and Related Problems

```
'''
Do P and P complement add up to 1?
Let us investigate
'''


import sympy
from sympy import FiniteSet

def probability(space, event):
    return len(event)/len(space)

def greaterThanFive(number):
    if number > 5:
        return True
    else:
        return False
def lessThanOrEqualToFive(number):
    if number <= 5:
        return True
    else:
        return False

if __name__ == '__main__':
    print(__doc__)
    space = FiniteSet(*range(1, 21))
    sixAndAbove = []
    fiveOrLess = []
    for num in space:
        if greaterThanFive(num):
            sixAndAbove.append(num)
    event= FiniteSet(*sixAndAbove)
    p = probability(space, event)
    for num in space:
```

```python
        if lessThanOrEqualToFive(num):
            fiveOrLess.append(num)
    eventComplement = FiniteSet(*fiveOrLess)
    pComplement = probability(space, eventComplement)

    print('Sample space: ',space)
    print('Events: ',sixAndAbove)
    print('EventComplement',fiveOrLess)
    print('Probability of Dice face being greater than 5: ',p)
    print('Probability of Dice face being 5 or less',pComplement)
    if ( p+pComplement == 1.0):
        print('Law of probability holds good')
    else:
        print('Things are incorrect')
```

```
Do P and P complement add up to 1?
Let us investigate

Sample space:  FiniteSet(1, 2, 3, 4, 5, 6, 7, 8, 9, 10, 11,
12, 13, 14, 15, 16, 17, 18, 19, 20)
Events:  [6, 7, 8, 9, 10, 11, 12, 13, 14, 15, 16, 17, 18, 19, 20]
EventComplement [1, 2, 3, 4, 5]
Probability of Dice face being greater than 5:  0.75
Probability of Dice face being 5 or less 0.25
Law of probability holds good
```

```python
'''
Demonstration of randomness of random.shuffle on list of objects
'''
import random

class Card:
    def __init__(self, suit, rank):
        self.suit = suit
        self.rank = rank
```

```python
def initializeDeck():
    suits = ['Clubs', 'Diamonds', 'Hearts', 'Spades']
    ranks = ['Ace', '2', '3','4', '5', '6', '7', '8', '9', '10', 'Jack', 'Queen', 'King']
    cards = []
    for suit in suits:
        for rank in ranks:
            card = Card(suit, rank)
            cards.append(card)
    return cards

def printSpecial(cards):
    for i in range(len(cards)):
        print('{',cards[i].suit,cards[i].rank,'}',end = "  ")
    print(' ')
    print('-----------------------------------------------------------')

def shuffleThenPrint(cards):
    random.shuffle(cards)
    printSpecial(cards)
    '''
    for card in cards:
        print('{0} of {1}'.format(card.rank, card.suit))
    '''

if __name__ == '__main__':
    print(__doc__)
    cards = initializeDeck()
    shuffleThenPrint(cards)
    shuffleThenPrint(cards)
    shuffleThenPrint(cards)
```

```
Demonstration of randomness of random.shuffle on list of objects
```

{ Clubs Jack } { Clubs King } { Spades 4 } { Clubs 5 } { Diamonds 4 } { Diamonds Jack } { Clubs 10 } { Hearts 9 } { Spades King } { Clubs 4 } { Hearts Queen } { Spades 3 } { Clubs 9 } { Hearts 3 } { Diamonds Ace } { Hearts 5 } { Spades Queen } { Hearts 4 } { Spades 5 } { Hearts 10 } { Diamonds 6 } { Hearts 7 } { Diamonds Queen } { Spades Jack } { Clubs Queen } { Diamonds 10 } { Spades 10 } { Hearts Jack } { Diamonds 3 } { Hearts 6 } { Clubs 7 } { Diamonds 7 } { Clubs Ace } { Diamonds 5 } { Clubs 2 } { Diamonds 9 } { Clubs 3 } { Hearts King } { Hearts 8 } { Spades Ace } { Diamonds 2 } { Spades 8 } { Clubs 8 } { Clubs 6 } { Spades 7 } { Hearts Ace } { Spades 6 } { Spades 9 } { Hearts 2 } { Diamonds King } { Diamonds 8 } { Spades 2 }

{ Clubs Queen } { Clubs 10 } { Clubs 2 } { Spades 9 } { Spades King } { Clubs King } { Hearts 8 } { Spades 10 } { Spades Jack } { Hearts Ace } { Hearts 7 } { Clubs 4 } { Hearts 6 } { Spades 2 } { Diamonds 6 } { Diamonds 3 } { Diamonds 2 } { Spades Queen } { Hearts King } { Hearts Queen } { Hearts 2 } { Spades 4 } { Clubs Ace } { Spades 3 } { Diamonds King } { Hearts 4 } { Clubs 3 } { Diamonds 10 } { Diamonds Queen } { Hearts 3 } { Spades 8 } { Spades 7 } { Diamonds 8 } { Diamonds 5 } { Hearts Jack } { Clubs Jack } { Diamonds Ace } { Spades Ace } { Diamonds 9 } { Hearts 10 } { Clubs 7 } { Clubs 9 } { Spades 6 } { Clubs 5 } { Clubs 6 } { Diamonds 7 } { Diamonds 4 } { Hearts 5 } { Spades 5 } { Clubs 8 } { Diamonds Jack } { Hearts 9 }

```
{ Spades 8 }    { Diamonds 6 }    { Clubs King }    { Spades 5 }   {
Diamonds King }    { Hearts 4 }    { Diamonds 8 }    { Diamonds Jack
}    { Spades 3 }    { Spades Queen }    { Clubs Ace }    { Diamonds
10 }    { Hearts 3 }    { Spades 7 }    { Clubs 2 }    { Hearts 8 }
{ Hearts 7 }    { Hearts 6 }    { Clubs 10 }    { Spades 2 }    {
Diamonds 7 }    { Hearts 2 }    { Clubs 7 }    { Clubs 9 }    { Clubs
Jack }    { Hearts 5 }    { Spades 4 }    { Clubs 6 }    { Diamonds 9
}    { Diamonds Queen }    { Hearts 10 }    { Hearts King }    {
Diamonds 2 }    { Spades King }    { Clubs 4 }    { Spades 9 }    {
Diamonds 3 }    { Hearts 9 }    { Spades Jack }    { Clubs 5 }    {
Diamonds 4 }    { Clubs 8 }    { Hearts Queen }    { Diamonds Ace }
{ Spades 6 }    { Spades Ace }    { Clubs Queen }    { Spades 10 }
{ Clubs 3 }    { Diamonds 5 }    { Hearts Ace }    { Hearts Jack }

----------------------------------------------------------------
```

```python
'''
Is numpy random randint is really random?
Let us visualize
'''

from mpl_toolkits import mplot3d
import numpy as np
import matplotlib.pyplot as plt

print(__doc__)
# Creating dataset
z = np.random.randint(80, size =(500))
x = np.random.randint(80, size =(500))
y = np.random.randint(80, size =(500))




# Creating figure
fig = plt.figure(figsize = (10, 7))
ax = plt.axes(projection ="3d")
```

```python
# Creating plot
ax.scatter3D(x, y, z, color = "green")
plt.title("simple 3D scatter plot")

# show plot
plt.show()
```

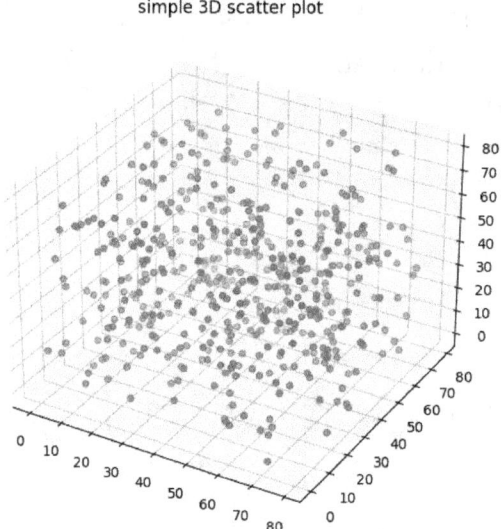

Figure 6: Random Points Generated by library

```python
'''
What effect mu and sigma have on probability distribution/density
function?
Let us investigate
'''


import matplotlib.pyplot as plt
import math
def normal_pdf(x, mu=0, sigma=1):
    sqrt_two_pi = math.sqrt(2 * math.pi)
```

```python
    return (math.exp(-(x-mu) ** 2 / 2 / sigma ** 2) /
(sqrt_two_pi * sigma))

xs = [x / 10.0 for x in range(-50, 50)]
plt.plot(xs,[normal_pdf(x,sigma=1) for x in xs],'-',label='mu=0,sigma=1')
plt.plot(xs,[normal_pdf(x,sigma=2) for x in xs],'--',label='mu=0,sigma=2')
plt.plot(xs,[normal_pdf(x,sigma=0.5) for x in xs],':',label='mu=0,sigma=0.5')
plt.plot(xs,[normal_pdf(x,mu=-1) for x in xs],'-.',label='mu=-1,sigma=1')
plt.legend()
plt.title("Various Normal pdfs")
plt.show()
```

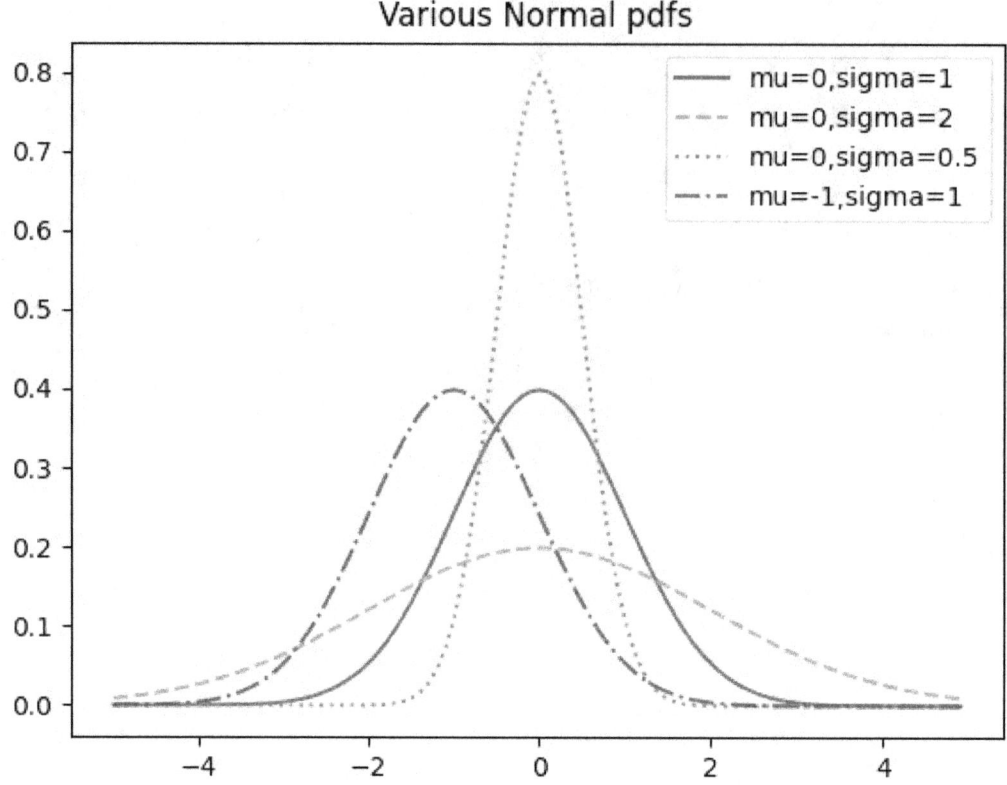

Figure 7: Normal Distribution for various mu and sigma

As sigma or standard deviation increases, the curve is more dispersed

```
'''
What effect mu and sigma have on cumulative distribution/density
function?
Let us investigate
'''

import matplotlib.pyplot as plt
import math

def normal_cdf(x, mu=0, sigma=1):
    return (1 + math.erf((x - mu) / math.sqrt(2) / sigma)) / 2
xs = [x / 10.0 for x in range(-50, 50)]
```

```python
plt.plot(xs,[normal_cdf(x,sigma=1) for x in xs],'-
',label='mu=0,sigma=1')
plt.plot(xs,[normal_cdf(x,sigma=2) for x in xs],'--
',label='mu=0,sigma=2')
plt.plot(xs,[normal_cdf(x,sigma=0.5) for x in
xs],':',label='mu=0,sigma=0.5')
plt.plot(xs,[normal_cdf(x,mu=-1) for x in
xs],'-.',label='mu=-1,sigma=1')
plt.legend(loc=4) # bottom right
plt.title("Various Normal cdfs")
plt.show()
```

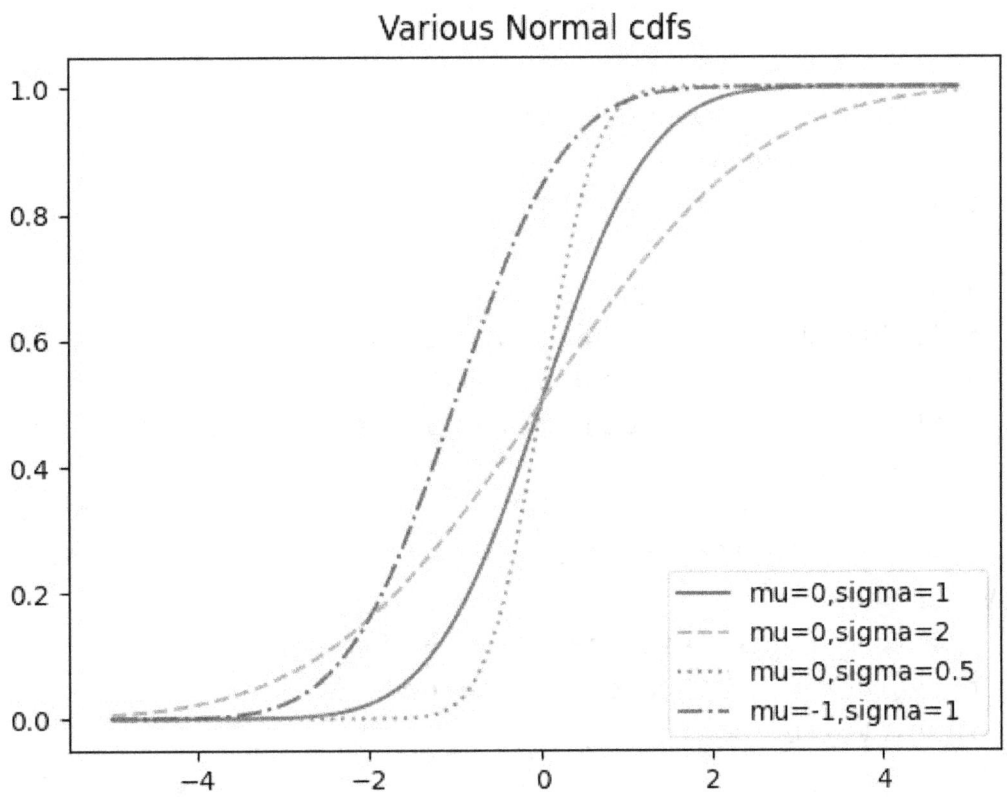

Figure 8: Various CDFs for different values of mu and sigma

> CDF is steep for lower values of sigma. Obvious. However mu value only shifts the CDF

```python
'''
In a class
10 students score 1
30 students score 2
30 students score 3
20 students score 4
10 students score 5
What is the expected value for this distribution and plot the
relevant plots
'''


import matplotlib.pyplot as plt
import math

def freqProbability(frequencyParam):
    print(frequencyParam)
    retList = []
    totalFrequency = 0
    for i in range(len(frequencyParam)):
        totalFrequency += frequencyParam[i]
    for i in range(len(frequencyParam)):
        retList.append(frequencyParam[i]/totalFrequency)
    return retList

def expectedValue(testScore, returnedList):
    expValue = 0
    for i in range(len(testScore)):
        expValue += testScore[i]*returnedList[i]
    return expValue

testScore = [1,2,3,4,5]
frequency = [10,30,30,20,10]

def mainMethod():
```

```
    returnedList = freqProbability(frequency)
    print('ExpectedValue ',expectedValue(testScore,returnedList))
    x_axis = list(set(testScore))
    plt.bar(x_axis, returnedList)
    plt.show()

mainMethod()
```

```
[10, 30, 30, 20, 10]
ExpectedValue  2.9
```

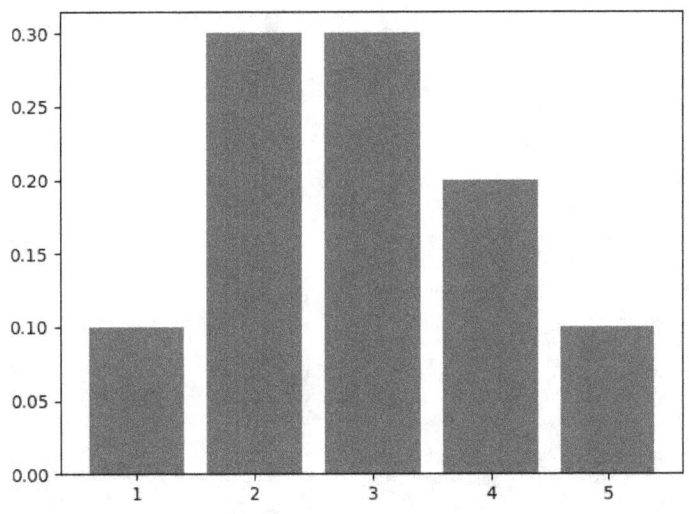

Figure 9: Distribution plot for the above problem

```
'''
Verify the accuracy of binomial library function with value
derived from first priciples
'''
```

```python
from sympy import FiniteSet
import random
from scipy.stats import binom

def findProb():
    coinSides = FiniteSet(0,1)
    maxTosses = 5
    successlist = []
    # sample space
    s = coinSides**maxTosses
    for elem in s:
        print(elem)
        if sum(elem) == 4:
            print('success')
            successlist.append(elem)
    e = FiniteSet(*successlist)
    print('Event space',e)
    print('Sample space',s)
    return len(e)/len(s)

if __name__ == '__main__':
    p = findProb()
    print('Probability from basic principles:   ',p)
    n = 5
    r = 4
    p = 0.5
    print('Probability with the help of library function:   ',binom.pmf(r,n,p))
```

```
(0, 0, 0, 0, 0)
(1, 0, 0, 0, 0)
(0, 1, 0, 0, 0)
(1, 1, 0, 0, 0)
(0, 0, 1, 0, 0)
(1, 0, 1, 0, 0)
(0, 1, 1, 0, 0)
```

```
(1, 1, 1, 0, 0)
(0, 0, 0, 1, 0)
(1, 0, 0, 1, 0)
(0, 1, 0, 1, 0)
(1, 1, 0, 1, 0)
(0, 0, 1, 1, 0)
(1, 0, 1, 1, 0)
(0, 1, 1, 1, 0)
(1, 1, 1, 1, 0)
success
(0, 0, 0, 0, 1)
(1, 0, 0, 0, 1)
(0, 1, 0, 0, 1)
(1, 1, 0, 0, 1)
(0, 0, 1, 0, 1)
(1, 0, 1, 0, 1)
(0, 1, 1, 0, 1)
(1, 1, 1, 0, 1)
success
(0, 0, 0, 1, 1)
(1, 0, 0, 1, 1)
(0, 1, 0, 1, 1)
(1, 1, 0, 1, 1)
success
(0, 0, 1, 1, 1)
(1, 0, 1, 1, 1)
success
(0, 1, 1, 1, 1)
success
(1, 1, 1, 1, 1)
Event space FiniteSet((0, 1, 1, 1, 1), (1, 0, 1, 1, 1), (1, 1, 0, 1, 1), (1, 1, 1, 0, 1), (1, 1, 1, 1, 0))
Sample space ProductSet(FiniteSet(0, 1), FiniteSet(0, 1), FiniteSet(0, 1), FiniteSet(0, 1), FiniteSet(0, 1))
Probability from basic principles:   0.15625
Probability with the help of library function: 0.15624999999999994
```

```python
'''
For given n and p values how does the pmf varies with r
Let us investigate
'''


from scipy.stats import binom
import matplotlib.pyplot as plt
n = 5
p = 0.5
# defining the list of r values
r_values = list(range(n + 1))
# obtaining the mean and variance
mean, var = binom.stats(n, p)
# list of pmf values
dist = [binom.pmf(r, n, p) for r in r_values ]
# printing the table
print("r\tp(r)")
for i in range(n + 1):
    print(str(r_values[i]) + "\t" + str(dist[i]))
# printing mean and variance
print("mean = "+str(mean))
print("variance = "+str(var))

dist = [binom.pmf(r, n, p) for r in r_values ]
# plotting the graph
plt.bar(r_values, dist)
plt.show()
```

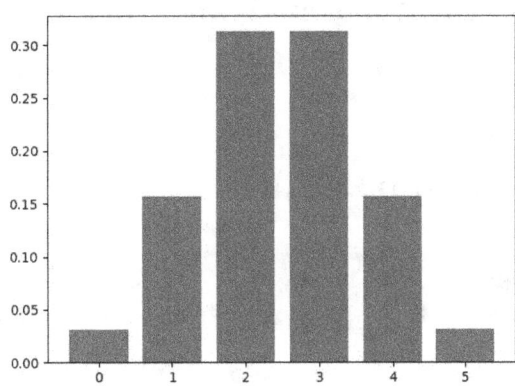

Figure 10: Binomial Equally likely

```
'''
Estimate the area of a circle
'''
import math
import random


# We are creating as many points as possible within square
# However, we are picking only those that are there within
# Circle. As we increase the points we will approach the
# area of circle
def estimate(radius, total_points):
    center = (radius, radius)

    in_circle = 0
    for i in range(total_points):
        x = random.uniform(0, 2*radius)
        y = random.uniform(0, 2*radius)
        p = (x, y)
        # distance of the point created from circle's center
        d = math.sqrt((p[0]-center[0])**2 + (p[1]-center[1])**2)
        if d <= radius:
            in_circle += 1
    area_of_square = (2*radius)**2
    print('In Circle',in_circle)
```

```python
        print('Total points',total_points)
        return (in_circle/total_points)*area_of_square

if __name__ == '__main__':
    radius = float(input('Radius: '))
    area_of_circle = math.pi*radius**2
    for points in [10**3, 10**5, 10**6]:
        print('Area: {0}, Estimated ({1}): {2}'.format(area_of_circle, points, estimate(radius, points)))
```

```
Radius: 5
In Circle 796
Total points 1000
Area: 78.53981633974483, Estimated (1000): 79.60000000000001
In Circle 78552
Total points 100000
Area: 78.53981633974483, Estimated (100000): 78.55199999999999
In Circle 785983
Total points 1000000
Area: 78.53981633974483, Estimated (1000000): 78.5983
Radius: 100
In Circle 772
Total points 1000
Area: 31415.926535897932, Estimated (1000): 30880.0
In Circle 78593
Total points 100000
Area: 31415.926535897932, Estimated (100000): 31437.2
In Circle 785096
Total points 1000000
Area: 31415.926535897932, Estimated (1000000): 31403.84
Radius: 1000
In Circle 771
Total points 1000
Area: 3141592.653589793, Estimated (1000): 3084000.0
In Circle 78612
Total points 100000
Area: 3141592.653589793, Estimated (100000): 3144480.0
```

```
In Circle 785352
Total points 1000000
Area: 3141592.653589793, Estimated (1000000): 3141408.0
```

Chapter 5 Statistics

Good Books:
1. Python for Probability Statistics and Machine Learning by Jose Unpingco
2. Think Stats Allen B. Downey
3. Practical Statistics for Data Scientists Peter Bruce, Andrew Bruce and Peter Gedeck
4. Statistics section in Scipy Lecture Notes.

Chapter 6 Scatter and Regression Models

In this section we demonstrate the Regression Models for Simple Linear Regression, Multiple Regression Analysis and Non Linear Regression

```python
import numpy as np
from sklearn.linear_model import LinearRegression
import matplotlib.pyplot as plt
x = np.array([5, 15, 25, 35, 45, 55]).reshape((-1, 1))
y = np.array([15, 35, 55, 75, 95, 115])
print(x)
print(y)
model = LinearRegression().fit(x, y)
r_sq = model.score(x, y)
plt.plot(x,y,'--k')
plt.plot(x, y, 'ok', ms=10)
plt.show()
print('r_squared:',r_sq)
print('intercept:', model.intercept_)
print('slope:', model.coef_)
y_pred = model.predict(x)
print('predicted response:', y_pred, sep='\n')
```

```
[[ 5]
 [15]
 [25]
 [35]
 [45]
 [55]]
[ 15  35  55  75  95 115]
r_squared: 1.0
intercept: 5.0
slope: [2.]
predicted response:
[ 15.  35.  55.  75.  95. 115.]
```

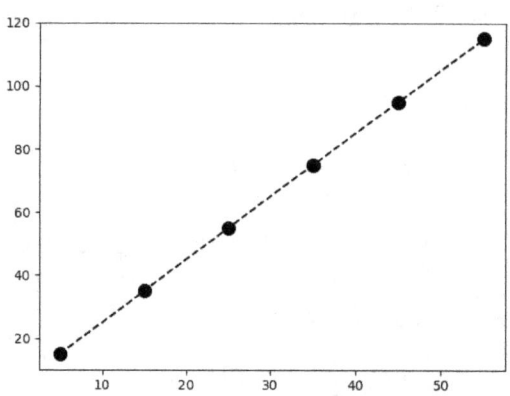

Figure 11: Simple Regression Model

```
import numpy as np
from sklearn.linear_model import LinearRegression
import matplotlib.pyplot as plt
x = np.array([5, 15, 25, 35, 45, 55]).reshape((-1, 1))
y = np.array([5, 20, 14, 32, 22, 38])
print(x)
print(y)
model = LinearRegression().fit(x, y)
r_sq = model.score(x, y)
plt.plot(x,y,'--k')
plt.plot(x, y, 'ok', ms=10)
print('r_squared:',r_sq)
print('intercept:', model.intercept_)
print('slope:', model.coef_)
y_pred = model.predict(x)
plt.plot(x,y_pred)
plt.show()
print('predicted response:', y_pred, sep='\n')
```

[[5]

```
 [15]
 [25]
 [35]
 [45]
 [55]]
[ 5 20 14 32 22 38]
r_squared: 0.715875613747954
intercept: 5.633333333333329
slope: [0.54]
predicted response:
[ 8.33333333 13.73333333 19.13333333 24.53333333 29.93333333
 35.33333333]
```

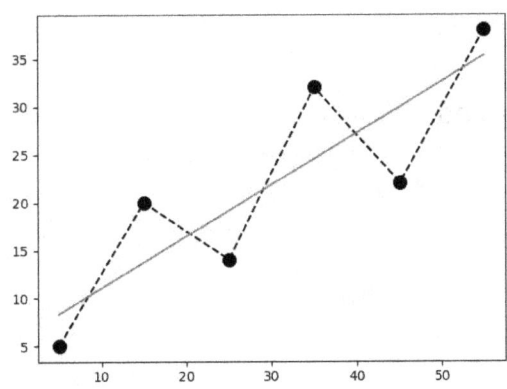

Figure 12: Simple Regression Model 2

```
import numpy as np
from sklearn.linear_model import LinearRegression
import matplotlib.pyplot as plt
from mpl_toolkits.mplot3d import Axes3D




ax = plt.axes(projection='3d')
```

```python
x = [[0, 1], [5, 1], [15, 2], [25, 5], [35, 11], [45, 15], [55, 34], [60, 35]]
y = [4, 5, 20, 14, 32, 22, 38, 43]
x, y = np.array(x), np.array(y)
x1 = x[...,0]
x2 = x[...,1]
print(x)
print(y)
print(x1)
print(x2)
ax.plot3D(x1, x2, y, 'gray')
model = LinearRegression().fit(x, y)
r_sq = model.score(x, y)
print('slope:', model.coef_)
print('r_squared:',r_sq)
print('intercept:', model.intercept_)
y_pred = model.predict(x)
ax.plot3D(x1, x2, y, 'red')
ax.plot3D(x1, x2, y_pred, 'blue')
print('predicted response:', y_pred, sep='\n')
plt.show()
```

```
[[ 0  1]
 [ 5  1]
 [15  2]
 [25  5]
 [35 11]
 [45 15]
 [55 34]
 [60 35]]
[ 4  5 20 14 32 22 38 43]
[ 0  5 15 25 35 45 55 60]
[ 1  1  2  5 11 15 34 35]
slope: [0.44706965 0.25502548]
r_squared: 0.8615939258756775
intercept: 5.52257927519819
predicted response:
```

```
[ 5.77760476  8.012953   12.73867497 17.9744479  23.97529728
 29.4660957
 38.78227633 41.27265006]
```

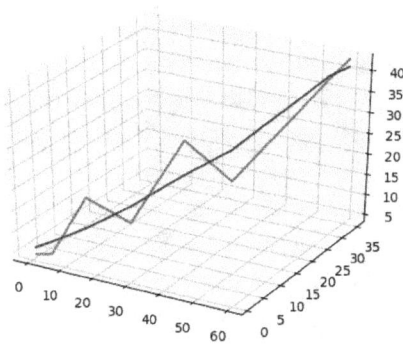

Figure 13: Multple Regression Model 3

```
import numpy as np
import matplotlib.pyplot as plt
from sklearn.linear_model import LinearRegression
from sklearn.preprocessing import PolynomialFeatures

x = np.array([5, 15, 25, 35, 45, 55]).reshape((-1, 1))
y = np.array([15, 11, 2, 8, 25, 32])
print(x)
transformer = PolynomialFeatures(degree=2, include_bias=False)
transformer.fit(x)
x_ = transformer.transform(x)
print(x_)
model = LinearRegression().fit(x_, y)
print(x_[...,1])
plt.plot(x_[...,1],y)

y_pred = model.predict(x_)
plt.plot(x_[...,1],y_pred)
plt.show()
```

```python
r_sq = model.score(x_, y)
print('r_squared:',r_sq)
print('intercept:', model.intercept_)
print('slope:', model.coef_)
```

```
[[ 5]
 [15]
 [25]
 [35]
 [45]
 [55]]
[[   5.   25.]
 [  15.  225.]
 [  25.  625.]
 [  35. 1225.]
 [  45. 2025.]
 [  55. 3025.]]
[  25.  225.  625. 1225. 2025. 3025.]
r_squared: 0.8908516262498564
intercept: 21.372321428571425
slope: [-1.32357143  0.02839286]
```

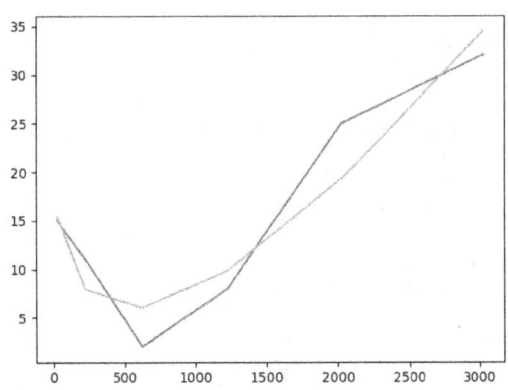

Figure 14: Non Linear Regression Model 4

```python
import numpy as np
import statsmodels.api as sm
x = [[1900, 30], [2000, 40], [1700, 30], [1400, 15], [1700, 32],
[1850, 38], [2000,27], [2300, 30],
[2200,26],[3752,35],[2300,18],[2500,17],[3800,40],[1740,12]]

y = [95000, 119000, 124800, 135000, 142800, 145000, 159000,
165000,182000,183000,200000,
211000,215000,219000]
x, y = np.array(x), np.array(y)
x = sm.add_constant(x)
print('x :' ,x)
print('y :',y)
model = sm.OLS(y, x)
results = model.fit()
print('regression coefficients:', results.params)
print(results.summary())
```

```
x : [[1.000e+00 1.900e+03 3.000e+01]
 [1.000e+00 2.000e+03 4.000e+01]
 [1.000e+00 1.700e+03 3.000e+01]
 [1.000e+00 1.400e+03 1.500e+01]
 [1.000e+00 1.700e+03 3.200e+01]
 [1.000e+00 1.850e+03 3.800e+01]
 [1.000e+00 2.000e+03 2.700e+01]
 [1.000e+00 2.300e+03 3.000e+01]
 [1.000e+00 2.200e+03 2.600e+01]
 [1.000e+00 3.752e+03 3.500e+01]
```

```
[1.000e+00 2.300e+03 1.800e+01]
[1.000e+00 2.500e+03 1.700e+01]
[1.000e+00 3.800e+03 4.000e+01]
[1.000e+00 1.740e+03 1.200e+01]]
y : [ 95000 119000 124800 135000 142800 145000 159000 165000 182000 183000
 200000 211000 215000 219000]
regression coefficients: [ 1.46740231e+05  4.36737528e+01 -2.86884933e+03]
/home/abhi/anaconda3/lib/python3.7/site-packages/scipy/stats/stats.py:1535: UserWarning: kurtosistest only
valid for n>=20 ... continuing anyway, n=14
  "anyway, n=%i" % int(n))
```

OLS Regression Results

Dep. Variable:	y	R-squared:	0.674
Model:	OLS	Adj. R-squared:	0.614
Method:	Least Squares	F-statistic:	11.36
Date:	Wed, 21 Apr 2021	Prob (F-statistic):	0.00211
Time:	19:16:10	Log-Likelihood:	-159.52
No. Observations:	14	AIC:	325.0
Df Residuals:	11	BIC:	327.0
Df Model:	2		
Covariance Type:	nonrobust		

	coef	std err	t	P>|t|	[0.025	0.975]
const	1.467e+05	2.54e+04	5.781	0.000	9.09e+04	2.03e+05
x1	43.6738	10.200	4.282	0.001	21.225	66.123
x2	-2868.8493	791.242	-3.626	0.004	-4610.362	-1127.337

Omnibus:	2.388	Durbin-Watson:	1.330
Prob(Omnibus):	0.303	Jarque-Bera (JB):	1.351
Skew:	-0.757	Prob(JB):	0.509
Kurtosis:	2.844	Cond. No.	9.13e+03

Notes:
[1] Standard Errors assume that the covariance matrix of the errors is correctly specified.
[2] The condition number is large, 9.13e+03. This might indicate that there are
strong multicollinearity or other numerical problems.

---

const and x1 and x2 coefficients and const are important

Simply $Y predicted = 146,700 + 43.6738 X1 - 2898.84 X2$

Problem definition from Qualitative Analysis for Management 11th Edition

Chapter 7 Forecasting

https://analyticsindiamag.com/time-series-forecasting-methods-in-python/
https://machinelearningmastery.com/time-series-forecasting-methods-in-python-cheat-sheet/

Chapter 8 Linear Programming and Applications

Plenty of online material is available including Scipy Ref Guide, Realpython, geeks for geeks and many more

We shall pick few cases from the Book Quantitative Analysis for Management 11th Edition and show few lines of Python code can do what

```
'''
Holiday meal turkey ranch page 290
'''

from scipy.optimize import linprog
obj = [2,3]

lhs_ineq = [[-5,-10 ],
            [-4,-3],
            [-0.5,0]]
rhs_ineq = [-90,
            -48,
            -1.5]
opt = linprog(c=obj, A_ub=lhs_ineq, b_ub=rhs_ineq,
              method="revised simplex")
print(opt)
```

```
     con: array([], dtype=float64)
     fun: 31.2
 message: 'Optimization terminated successfully.'
     nit: 2
   slack: array([0. , 0. , 2.7])
  status: 0
 success: True
       x: array([8.4, 4.8])
```

```
from scipy.optimize import linprog
'''
```

```
Media selection page number 328
'''
obj = [-5000, -8500,-2400,-2800]

lhs_ineq = [[ 1,0,0,0],
            [0,1,0,0],
            [0,0,1,0],
            [0,0,0,1],
            [800,925,290,380],
            [0,0,-1,-1],
            [0,0,290,380]]

rhs_ineq = [12,
            5,
            25,
            20,
            8000,
            -5,
            1800]
opt = linprog(c=obj, A_ub=lhs_ineq, b_ub=rhs_ineq,
              method="revised simplex")
print(opt)
```

```
     con: array([], dtype=float64)
     fun: -67240.30172413793
 message: 'Optimization terminated successfully.'
     nit: 4
   slack: array([10.03125   ,  0.        , 18.79310345, 20.
,  0.        ,
          1.20689655,  0.        ])
  status: 0
 success: True
       x: array([1.96875   , 5.        , 6.20689655, 0.        ])
```

```
from scipy.optimize import linprog
'''
```

```
Manufacturing application problem page number 332
'''

obj = [-16.24, -8.22, -8.77, -8.66]

lhs_ineq = [[0.125, 0, 0, 0.066],
            [0, 0.08, 0.05, 0],
            [0, 0, 0.05, 0.044]]

rhs_ineq = [ 1200,
             3000,
             1600]

bnd = [(5000, 7000),
       (10000,14000),
       (13000,16000),
       (5000, 8500)]

opt = linprog(c=obj, A_ub=lhs_ineq, b_ub=rhs_ineq, bounds=bnd,
              method="revised simplex")
print(opt)
```

```
     con: array([], dtype=float64)
     fun: -412028.88
 message: 'Optimization terminated successfully.'
     nit: 4
   slack: array([   0., 1080.,  426.])
  status: 0
 success: True
       x: array([ 5112., 14000., 16000.,  8500.])
```

We picked few Random examples from Book Quantitative Analysis for Management and demonstrated the solution with Python

Chapter 9 Introduction to Basic Sympy Functionalities

We shall demonstrate the capability of Sympy through small code snippets

Demonstrate the creation of Symbols and print some algebraic operations

1

```python
#!/usr/bin/env python

"""Basic example

Prepared with help of Sympy documentation

Demonstrates how to create symbols and print some algebra
operations. Demonstrates the pprint aka PrettyPrint
"""

from sympy import Symbol, pprint


def main():
    print(__doc__) #Print the documentation
    a = Symbol('a')
    b = Symbol('b')
    c = Symbol('c')
    e = ( a*b*b + 2*b*a*b )**c

    print('')
    pprint(e)    #Solves the expression and does pretty printing
    print('')

if __name__ == "__main__":
    main()
```

```
$ python basic_chap02.py
Basic example

Prepared with help of Sympy documentation
```

Demonstrates how to create symbols and print some algebra operations. Demonstrates the pprint aka PrettyPrint

$$\left(3 \cdot a \cdot b^2\right)^c$$

2

Demonstrate the Differentiation with Sympy

```
#!/usr/bin/env python

"""Differentiation example

Demonstrates some differentiation operations.
"""

from sympy import pprint, Symbol

def main():
    a = Symbol('a')
    b = Symbol('b')
    e = (a + 3*b)**6

    print("\nExpression : ")
    print()
    pprint(e)
    print("\n\nDifferentiating w.r.t. a:")
    print()
    pprint(e.diff(a))
    print("\n\nDifferentiating w.r.t. b:")
    print()
    pprint(e.diff(b))
    print("\n\nSecond derivative of the above result w.r.t. a:")
    print()
    pprint(e.diff(b).diff(a, 2))
    print("\n\nExpanding the above result:")
```

```
        print()
        pprint(e.expand().diff(b).diff(a, 2))
        print()

if __name__ == "__main__":
    main()
```

```
$ python differentiation_chap02.py

Expression :

          6
(a + 3·b)


Differentiating w.r.t. a:

          5
6·(a + 3·b)


Differentiating w.r.t. b:

           5
18·(a + 3·b)


Second derivative of the above result w.r.t. a:

           3
360·(a + 3·b)


Expanding the above result:

     ⎛ 3       2        2       3⎞
360·⎝a  + 9·a ·b + 27·a·b  + 27·b ⎠
```

Demonstrate the Algebraic expression expansion capability with Sympy

```python
#!/usr/bin/env python

""" Algebraic Expression Expansion Example

Demonstrates how to expand expressions.

Prepared with the help of Sympy documentation
"""

from sympy import pprint, Symbol

def main():
    print(__doc__)
    a = Symbol('a')
    b = Symbol('b')
    c = (a + b)**2
    d = (a +b)**5
    e = (a + b)**10

    print("\nExpression:")
    pprint(c)
    print('\nExpansion of the above expression:')
    pprint(c.expand())
    print("\nExpression:")
    pprint(d)
    print('\nExpansion of the above expression:')
    pprint(d.expand())
    print("\nExpression:")
    pprint(e)
    print('\nExpansion of the above expression:')
    pprint(e.expand())
    print()

if __name__ == "__main__":
```

```
    main()
```

```
$python expansion_chap02.py
 Algebraic Expression Expansion Example

Demonstrates how to expand expressions.

Prepared with the help of Sympy documentation


Expression:
        2
(a + b)

Expansion of the above expression:
 2           2
a  + 2·a·b + b

Expression:
        5
(a + b)

Expansion of the above expression:
 5      4        3  2       2  3      4     5
a  + 5·a ·b + 10·a ·b  + 10·a ·b  + 5·a·b  + b

Expression:
        10
(a + b)

Expansion of the above expression:
 10       9        8  2        7  3        6  4        5  5
a   + 10·a ·b + 45·a ·b  + 120·a ·b  + 210·a ·b  + 252·a ·b  +
     4  6        3  7       2  8        9     10
210·a ·b  + 120·a ·b  + 45·a ·b  + 10·a·b  + b
```

Demonstrate the Mathematics functions with Sympy

```python
#!/usr/bin/env python

"""Functions example

Demonstrates functions defined in SymPy.

Prepared with help of Sympy documentation
"""

from sympy import pprint, Symbol, log, exp

def main():
    print(__doc__)
    a = Symbol('a')
    b = Symbol('b')
    e = log((a + b)**5)
    print('Constructed expression is')
    pprint(e)
    print('\n')

    print(' e raise to the expression is')
    e = exp(e)
    pprint(e)
    print('\n')

    print('Another constructed expression is')
    e = log(exp((a + b)**5))
    pprint(e)
    print

if __name__ == "__main__":
    main()
```

```
$python functions_chap02.py
Functions example

Demonstrates functions defined in SymPy.

Prepared with help of Sympy documentation

Constructed expression is
     ⎛        5⎞
  log⎝(a + b) ⎠


 e raise to the expression is
         5
  (a + b)


Another constructed expression is
     ⎛ ⎛        5⎞⎞
     ⎜ ⎝(a + b) ⎠⎟
  log⎝e          ⎠
```

Demonstration of Limits with the help of Sympy

```
#!/usr/bin/env python

"""Limits Example

Demonstrates limits.

Sympy documentation team
"""

from sympy import exp, log, Symbol, Rational, sin, limit, sqrt, oo


def sqrt3(x):
```

```python
        return x**Rational(1, 3)


def show(computed, correct):
    print("computed:", computed, "correct:", correct)


def main():
    print(__doc__)
    x = Symbol("x")

    show( limit(sqrt(x**2 - 5*x + 6) - x, x, oo), -Rational(5)/2 )

    show( limit(x*(sqrt(x**2 + 1) - x), x, oo), Rational(1)/2 )

    show( limit(x - sqrt3(x**3 - 1), x, oo), Rational(0) )

    show( limit(log(1 + exp(x))/x, x, -oo), Rational(0) )

    show( limit(log(1 + exp(x))/x, x, oo), Rational(1) )

    show( limit(sin(3*x)/x, x, 0), Rational(3) )

    show( limit(sin(5*x)/sin(2*x), x, 0), Rational(5)/2 )

    show( limit(((x - 1)/(x + 1))**x, x, oo), exp(-2))

if __name__ == "__main__":
    main()
```

```
$ python limits_examples_chap02.py
Limits Example

Demonstrates limits.

Sympy documentation team
```

```
computed: -5/2 correct: -5/2
computed: 1/2 correct: 1/2
computed: 0 correct: 0
computed: 0 correct: 0
computed: 1 correct: 1
computed: 3 correct: 3
computed: 5/2 correct: 5/2
computed: exp(-2) correct: exp(-2)
```

Demonstration of Arbitrary Precision with Sympy

```python
#!/usr/bin/env python

"""Precision Example

Demonstrates SymPy's arbitrary integer precision abilities

Prepared with the help of Sympy Documentation
"""


import sympy
from sympy import Mul, Pow, S, pprint


def main():
    print(__doc__)
    x = Pow(2, 50, evaluate=False)
    y = Pow(10, -50, evaluate=False)
    # A large, unevaluated expression
    m = Mul(x, y, evaluate=False)
    # Evaluating the expression
    print('Value of Expression')
    e = S(2)**50/S(10)**50
    pprint(m)
    print('is as follows:')
    print(e)
if __name__ == "__main__":
    main()
```

```
$ python precision_chap02.py
Precision Example

Demonstrates SymPy's arbitrary integer precision abilities

Prepared with the help of Sympy Documentation

Value of Expression
 50
2
─────
  50
10
is as follows:
1/88817841970012523233890533447265625
```

Expression evaluation and printing with the help of pretty print

```
#!/usr/bin/env python

"""Pretty print example

Demonstrates pretty printing.

Refer Sympy documentation for more details
"""

from sympy import Symbol, pprint, sin, cos, exp, sqrt,
MatrixSymbol, KroneckerProduct, Matrix


def main():

    print(__doc__)
```

```python
x = Symbol("x")
y = Symbol("y")



a = MatrixSymbol("a", 1, 1)
b = MatrixSymbol("b", 1, 1)
c = MatrixSymbol("c", 1, 1)

pprint( x**x )
print('\n')   # separate with two blank likes

pprint(x**2 + y + x)
print('\n')

pprint(sin(x)**x)
print('\n')

pprint( sin(x)**cos(x) )
print('\n')

pprint( sin(x)/(cos(x)**2 * x**x + (2*y)) )
print('\n')

pprint( sin(x**2 + exp(x)) )
print('\n')

pprint( sqrt(exp(x)) )
print('\n')

pprint( sqrt(sqrt(exp(x))) )
print('\n')

pprint( (1/cos(x)).series(x, 0, 10) )
print('\n')

A= Matrix([[1, -1, 1], [3, 4, 5], [0, 2, 6]])
pprint(A)
print('\n')
```

```
        pprint(a*(KroneckerProduct(b, c)))
        print('\n')

if __name__ == "__main__":
    main()
```

```
$ python print_pretty_chap02.py
Pretty print example

Demonstrates pretty printing.

Refer Sympy documentation for more details

 x
x


 2
x  + x + y


    x
sin (x)


    cos(x)
sin    (x)


     sin(x)
    ─────────
     x    2
    x ·cos (x) + 2·y


    ⎛ 2    x⎞
```

$$\sin\left(x + e^x\right)$$

$$\sqrt{e^x}$$

$$\sqrt[4]{e^x}$$

$$1 + \frac{x^2}{2} + \frac{5 \cdot x^4}{24} + \frac{61 \cdot x^6}{720} + \frac{277 \cdot x^8}{8064} + O\left(x^{10}\right)$$

$$\begin{bmatrix} 1 & -1 & 1 \\ 3 & 4 & 5 \\ 0 & 2 & 6 \end{bmatrix}$$

$$a \cdot (b \otimes c)$$

Demonstration of series capability with Sympy

```
#!/usr/bin/env python

"""Series example
```

```
Demonstrates series.

Prepared with the help of sympy documentation
"""

from sympy import Symbol, cos, sin, pprint


def main():
    print(__doc__)
    x = Symbol('x')

    e = 1/cos(x)
    print('')
    print("Series for sec(x):")
    print('')
    pprint(e.series(x, 0, 10))
    print("\n")

    e = 1/sin(x)
    print("Series for csc(x):")
    print('')
    pprint(e.series(x, 0, 4))
    print('')

if __name__ == "__main__":
    main()
```

```
$python series_chap02.py
Series example


Demonstrates series.

Prepared with the help of sympy documentation


Series for sec(x):
```

$$1 + \frac{x^2}{2} + \frac{5 \cdot x^4}{24} + \frac{61 \cdot x^6}{720} + \frac{277 \cdot x^8}{8064} + O\left(x^{10}\right)$$

Series for csc(x):

$$\frac{1}{x} + \frac{x}{6} + \frac{7 \cdot x^3}{360} + O\left(x^4\right)$$

Demonstration of Substitution capability of Sympy

```python
#!/usr/bin/env python

"""Substitution example

Demonstrates substitution.
"""

import sympy
from sympy import pprint, cos, Float, Symbol, log


def main():
    x = Symbol('x')
    y = Symbol('y')

    print('Let us start with')
    e = 1/cos(x)
    print()
    pprint(e)
    print('\n')
```

```python
    print('Let us substitute cos(x) with y')
    pprint(e.subs(cos(x), y))
    print('\n')
    print('Let us substitute y with x square')
    pprint(e.subs(cos(x), y).subs(y, x**2))   #Watch the cascading of subs

    e = 1/log(x)
    print('Given expression')
    pprint(e)
    print('Substitute 2.71828 for x')
    e = e.subs(x, sympy.Float("2.71828"))
    print('\n')
    print('Evaluated expression is ')
    pprint(e.evalf())
    print()

    print('Let us construct following expression')
    a = Symbol('a')
    b = Symbol('b')
    e = a*2 + a**b/a
    print('\n')
    pprint(e)

    print('Substitute a with 8')
    print('\n')
    pprint(e.subs(a,8))
    print()

if __name__ == "__main__":
    main()
```

```
$ python substitution_chap02.py
Let us start with

  1
```

$$\cos(x)$$

Let us substitute cos(x) with y

$$1 - y$$

Let us substitute y with x square

$$\frac{1}{x^2}$$

Given expression

$$\frac{1}{\log(x)}$$

Substitute 2.71828 for x

Evaluated expression is
1.00000067265317

Let us construct following expression

$$2 \cdot a + \frac{a^b}{a}$$

Substitute a with 8

$$\frac{8^b}{8} + 16$$

Summary:

We demonstrated the ability of Sympy with the help of examples. The examples are again based on the Sympydoc.

Chapter 10 Plotting with Sympy

Have a look at the following code block

```
from sympy import *
from sympy.plotting import *

x,y = symbols('x y')

p = plot3d_parametric_surface(sin(x)*sin(y), sin(x)*cos(y),
cos(x), (x, 0, pi), (y, 0, 2*pi))

p = plot3d(sin(x)*y, (x, 0, 6*pi), (y, -5, 5))
```

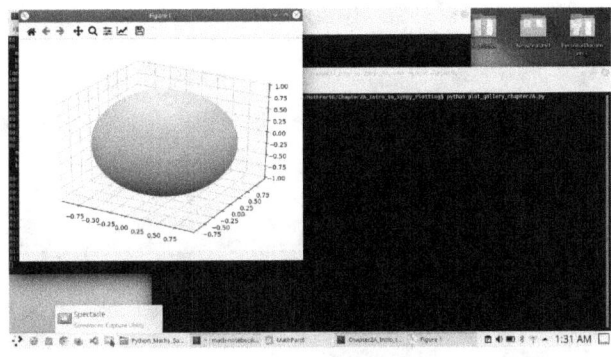

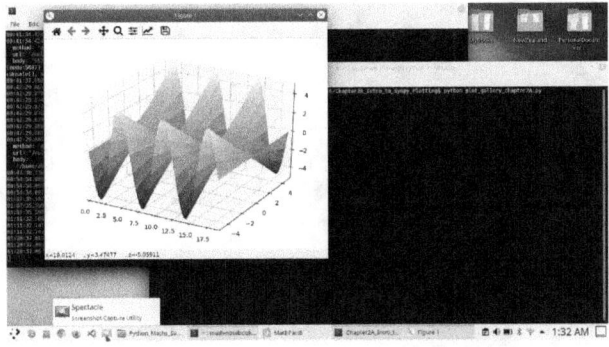

While sin(x)*y is spreading the sin(x) wave with y when x takes on values 0 to 6pi and y takes the value -5 to 5, The parametric surface is not so easy to imagine without computers.

Chapter 11 Summary

We touched the tip of ice berg by taking few cases of Quantitative methods with Python. There is a high probability that you will find some one has already documented their own experiences with python in the form Books, Blogs etc. Today the Scipy and Sympy user guides combined is in excess of 5000 pages. That is enormous capability available to everyone to carry out their activity in Quantitative Field with Python. And it is growing by the day. It is these libraries that make Python so Rich!!